STAR GAZE

BY ANDRE CLARKE

This is a work of fiction, and the views expressed herein are the sole responsibility of the author. Likewise, certain characters, places, and incidents are the products of the author's imagination, and any resemblance to actual persons, living or dead, or actual events or locales, is entirely coincidental.

Star Gaze (Book Series, Book 1)

Published by Ateem Marketing LLC

460 East Fordham Rd

Bronx, New York 10458

©2021 Andre Clarke www.dreclarke.com

Illustrated by Aquil Khan ©

All rights reserved. No parts of this book may be scanned, uploaded, reproduced, distributed, or transmitted in any form or by any means whatsoever without written permission from the author, except in the case of brief quotations embodied in critical articles and reviews. Thank you for supporting the author's rights.

Published 2021

Printed in the United States of America

ISBN: 978-1-7364622-0-1

Acknowledgment

The Ateam; Andrea, Arlo, Aria, the real shining stars in my life!

My family, thank you for your help and support(Mom, Rohan, Keisha, Adrian, Richard, Robin and Thio)

My nephews of course. Zachary, Jaedyn, Aiden, Athan, Laiken, Tyler and my nieces Raegan and Yanique.

Everyone helped me get this book completed. Earl and Yvie, Dave and Charlene, Troy and Melissa, Janet, Websters, and Heslop. I am grateful for everything.

Thank you to the staff at lola Jordan Daycare. (Ms. Mendez, Ms. Flores, Ms. Wynn, Ms. Salgado, Ms. Benny, Ms. Boxhill, and many others). Princeton University.

The Hunter Hawks staff at KIES. Ms. Germosen and Ms. Hiskey and many others.

The gift of every child is to dream big. Such is the case for young Zachary, a child with a big heart and an even bigger imagination. Star Gaze is the story of how a child's passion and ingenuity translates into a meteoric event, one that might change his life forever. Journey with our young explorer as he goes from student to teacher and turns his dream of playing amongst the stars into reality.

Star Gaze will surely become a favorite of-any child's literary collection.

Zachary was a very smart boy who loved astronomy. He often thought of a life beyond Earth and would make up stories about aliens, star animals, and new planets.

"I bet there's a Jamaican alien out there too!" He said to his parents, who laughed.

"Of course, Jamaicans live everywhere," his Dad replied jovially.

"So, what do you want for your birthday, Love?" His Mom asked.

"A telescope, a real one!" He exclaimed.

"We'll try our best son," his Dad said. But noticing their exchanged glance, Zachary realized his parents couldn't afford it. So he told them he wanted the new Alien on Earth virtual video game. Something he knew they could afford.

That night, Zachary had a bizarre dream. He dreamed a star spoke to him—Sirius, to be exact.

"Zachary, you can build it. Build your telescope," the star said. He drifted back to sleep, with the words of Sirius filling his mind.

Zachary jumped out of bed the next morning excited, "Yes! I'm going to build it"!

After breakfast, he began researching telescope construction. While doing his research, he soon found out that to build one, he would have to buy parts.

He looked at the prices of the parts and said, "Whew, that's like a gazillion dollars to me."

There was no way that he could save adequately to buy the parts, so Zachary racked his brain for money-making ideas. But for a child his age, generating income would not be easy.

Zachary fell asleep that night with money on his mind.

The next morning at school, Zachary asked his classmates and teachers if they had any ideas on how he could make money. They gave him many suggestions, but he would probably be his parent's age by the time he would save enough.

"What's wrong, Zachary?" His Tech teacher asked him when she noticed he was unusually quiet. Zachary explained to Ms. Benny what he was trying to do, and she gave him a brilliant idea: apps! She believed he could do it and told him so.

Zachary went home that day wondering what kind of app he could create to help him make money and get people's attention at the same time. Then it came to him.

The next morning was the start of a long holiday weekend. Zachary woke up earlier than usual, and he had a lot of time before he went back to school. He already knew enough about creating apps because they fascinated him. Zachary excelled in Tech.

He started coding on his computer- as fast as he could. His fingers flying, he was so caught up with coding that he didn't realize it was now nighttime. Zachary knew he had to get some sleep as his brain worked best when he felt rested. So he turned off the computer, then got ready for bed.

Zachary wasn't sure when he fell asleep, but he remembers looking up at the stars.

The next day, he was at it again.

"What are you up to, honey?" His mother asked while his Dad listened in. They had never seen Zachary so focused. He did not touch his video games or played all weekend.

"It's a starprise, Mom." He winked. And the next day he was at it again, building and testing. Zachary knew it sounded impossible to build an app in such a short period, but he remembered his parents telling him that if he could perceive it, he could achieve it!

Finally, he completed work on the prototype.

"Mom, Dad!" He called them excitedly early on his birthday morning.

"Zachary, you're up early! Are you okay?" They rushed into his room.

"Yes, yes! I know every year you've asked me what I want for my birthday, and I tell you something expensive and then take it back. I'm sorry."

"No, son. It's fine. You have big dreams, and you have to pursue them," his Dad said.

"This year, I made something as my gift to you, for us. I present to you, Star Gaze!" He showed them the app. Where they could see the stars up close.

"Wow, it's like a telescope on your phone," his Mom said.

"Yes!" declared Zachary.

"Zachary, we are so proud of you!" His parents hugged him tightly. Soon after, he sent Ms. Benny the link to his app. She shared it with several others, and by 1 p.m. that day, Zachary received a life-changing call. Well, his Mom picked up the phone.

"Um Zachary, someone wants to invest in your app!" She hugged him again as she shared the good news.

"Wow," Zachary couldn't believe it. The investors would meet with Zachary and his parents that Friday to discuss details. After they completed the call, Zachary and his parents celebrated. They screamed, danced, jumped, and cried tears of joy.

"Now, we can get that house we always wanted," Zachary said and hugged his parents.

"But first open your present," his parents said and gave him a heavy gift... He unwrapped it faster than you could say his name and there it was a telescope! A handmade telescope with a note from his parents—one day we will get you a real one, son. Love, Mom, and Dad.

Zachary gazed at his parents and realized that they were the true and permanent stars in his Life.

Glossary

Astronomy — science that encompasses the study of all objects in space (star, planets)

Sirius — brightest star in space

Telescope — an instrument designed to make distant objects appear nearer

Virtual — not physically existing as such but made by software to appear to do so

Invest — to put money, effort, time, etc. into something to make a profit or get an advantage

Perceive — come to realize or understand.

Achieve — to accomplish something

Prototype — a first, typical or preliminary model of something

Coding — process of writing computer software code

Tech — area of work that does or makes something in technology

Join Zachary on His Amazing Adventure That Will Inspire Every Child to Work Hard and Dream Big.

Would you like to:

- Get an idea of how to make your dreams come true?
- Spark your imagination?
- Explore alternative possibilities?

If so, you're in luck!

We have just the right motivational storybook for you to enjoy.

Meet Zachary. He loves astronomy and would often imagine aliens, stars, and new planets.

His birthday is coming up soon, and he asked his parents for a telescope. In a twist of fate, Zachary had a dream that would change his life.

Brimming with energy, Zachary gets creative and takes matters into his own hands.

Would Zachary be able to achieve his goal, or would he give up?

Find out and join Zachary on his adventure as he begins his journey as a creator and turns his dream of playing amongst the stars into reality.

Enjoy this inspiring tale, aiming to show every child how to achieve their goals while exploring the areas of entrepreneurship, coding, and tech.

Here's what this book will offer you:

- **Empowering story:** Hop on a journey that will inspire you to dream big and make your dreams come true.

- **Self-confidence:** Instill determination to help you try new things, explore challenges, and solve problems.

- **Achievement guide:** Inspire your mind to discover the possibilities you can achieve as long you work hard for it.

- **Career Guide:** Perfect introduction for kids to entrepreneurship, coding, and tech.

- **Culture of teamwork:** Learn how asking for help and working as a team can help you achieve your goals faster

And so much more.

If you want to spark your imagination, explore ideas, and take on new challenges, this is the perfect adventure book for you. Join Zachary and make your dreams become reality.

Made in the USA
Monee, IL
30 January 2021